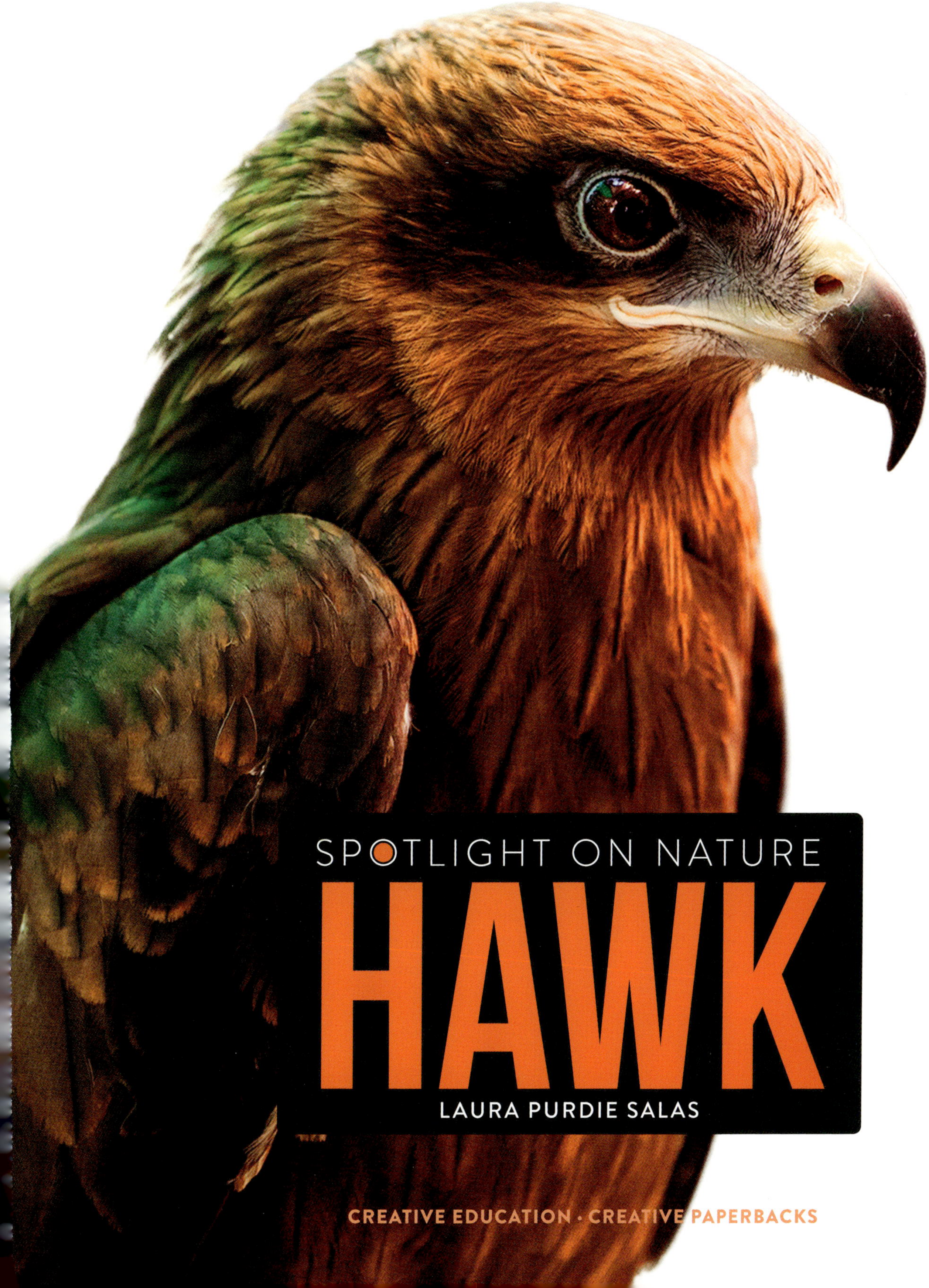

SPOTLIGHT ON NATURE

HAWK

LAURA PURDIE SALAS

CREATIVE EDUCATION · CREATIVE PAPERBACKS

Published by Creative Education and Creative Paperbacks
P.O. Box 227, Mankato, Minnesota 56002
Creative Education and Creative Paperbacks are imprints
of The Creative Company
www.thecreativecompany.us

Design by Rhea Magaro; production by Beeline Media and Design, Inc.
Art direction by Tom Morgan
Printed in the United States of America

Photographs by Alamy (Roy Waller, MediaMedium), Dreamstime (Pavel Mikoska), Getty Images (mallardg500, Carol Gray), Shutterstock (apple2499, Yulik_art, CptNaveen, Gallinago_media, arsal1, Henk Bogaard, Honyojima, Pictureguy, Volodymyr Burdiak, Labrynthe, Chase D'animulls, Rob Palmer Photography, Marcin Perkowski, James Arup, Thomas Barrat, Lucy Jorna, J. S. Fisher, Elliotte Rusty Harold)

Library of Congress Cataloging-in-Publication Data
Names: Salas, Laura Purdie, author.
Title: Hawk / Laura Purdie Salas.
Description: Mankato, Minnesota : Creative Education and Creative Paperbacks, [2025] | Series: Spotlight on nature | Includes bibliographical references and index. | Audience: Ages 10-13 | Audience: Grades 4-6 | Summary: "Soar into the hawk's world with our new book for mid-level readers. An engaging narrative of a hawk family, captivating images, infographics, milestones, and resources make it perfect for young nature enthusiasts"-- Provided by publisher.
Identifiers: LCCN 2023059461 (print) | LCCN 2023059462 (ebook) | ISBN 9798889891932 (library binding) | ISBN 9781682775660 (paperback) | ISBN 9798889892014 (ebook)
Subjects: LCSH: Red-tailed hawk--Juvenile literature. | Red-tailed hawk--Life cycles--Juvenile literature. | Hawks--Juvenile literature. | Hawks--Life cycles--Juvenile literature. | Red-tailed hawk--Sonoran Desert--Juvenile literature.
Classification: LCC QL696.F32 S248 2025 (print) | LCC QL696.F32 (ebook) |
DDC 598.9/44--dc23/eng/20240123
LC record available at https://lccn.loc.gov/2023059461
LC ebook record available at https://lccn.loc.gov/2023059462

CONTENTS

MEET THE FAMILY

Red-Tailed Hawks of the Sonoran Desert 4

LIFE BEGINS 7

Featured Family

Welcome to the World 8

First Meal 10

EARLY ADVENTURES 13

Featured Family

Look Who's Jump-Flapping 14

Give It a Try 16

LIFE LESSONS 19

Featured Family

This Is How It's Done 20

Practice Makes Perfect 22

HELPING HAWKS SURVIVE 25

Family Album Snapshots 28

Words to Know 30

Learn More 31

Index 32

MEET THE FAMILY

RED-TAILED HAWKS

of the Sonoran Desert

It is May in the rugged Sonoran Desert. This region of the southwestern United States is dry and dusty. The blazing sun heats the afternoon. Bushes and tall cactuses dot the landscape. This desert looks rather empty, but it's not. More than 2,000 plant **species** and around 560 animal species live here. One plant, the saguaro cactus, is tall and spiny. It doesn't look like it would make a comfortable home.

However, a pair of red-tailed hawks has built a nest high in this cactus. It is a sturdy structure built from sticks. From here, the hawks can view the land for miles around. The mother keeps the three eggs warm while the father patrols the skies. He dives from the nest, stretching out his **talons**. Soon afterward, he flies a dead rabbit to the nest to feed his mate. The hawks have spent the past month following this routine. But today, everything will change. The eggs are about to hatch.

CLOSE-UP

Beak

A hawk's beak has a bony core. The core is covered with keratin, the same protein that forms human fingernails. The beak has a sharp, curved hook that makes it a good tool for slicing flesh.

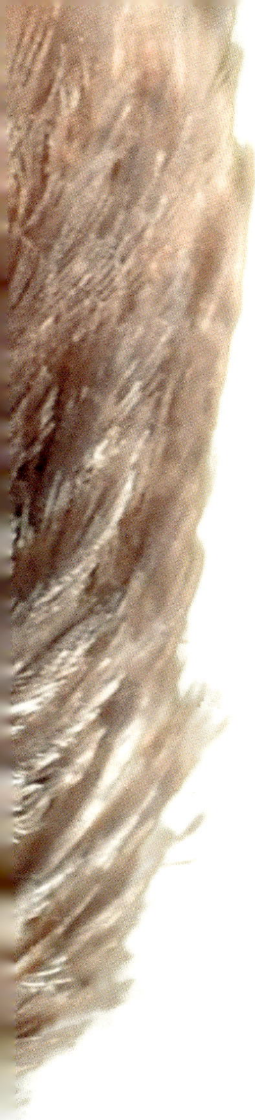

CHAPTER ONE

LIFE BEGINS

Hawks are medium-sized **raptors**, or birds of prey. Hawks generally include two groups of birds: **accipiters** and **buteos**. Accipiters are also called true hawks. They are lean and have short, rounded wings and long tails. When they fly overhead, they almost look like a cross. Buteos are stockier hawks. They have broad wings and wide tails.

All hawks have sharp beaks and talons along with excellent eyesight. Their feathers can be black, gray, white, brown, gold, or rusty red. Different hawk species have different colors and patterns. They also grow to different sizes. A fully grown sharp-shinned hawk weighs only seven ounces (198 grams). Its wingspan is about 22 inches (56 centimeters). The ferruginous hawk is much larger. It can weigh almost five pounds (two kilograms). Its wingspan stretches more than 4.5 feet (1.4 meters).

RED-TAILED HAWK MILESTONES

DAY 1

- Hatches from egg
- Weight: 2 ounces (58 g)
- Has fluffy, fuzzy, white feathers
- Weak and clumsy
- Gets fed beak-to-beak by parents

Hawks live on every continent except Antarctica. True hawks, including most sparrowhawks and goshawks, live mainly in forests. They prey mostly on smaller birds, but they also eat some small mammals. Buteos, such as red-tailed hawks and Swainson's hawks, prefer open lands. They're found in deserts, prairies, and grasslands. Their diet depends on where they live. They might eat small mammals, birds, reptiles, and insects.

CLOSE-UP

Talons

Talons are razor-sharp claws. Hawks use them to catch, hold, and kill prey. Each foot has three talons that face forward and one talon that faces backward.

FEATURED FAMILY

Welcome to the World

The hawk eggs rest in the nest's inner cup. This cup is lined with soft strips of bark and leaves. Now, a crack appears in one of the eggs. The mother hawk moves about fussily. She uses her sharp beak and talons to carefully rotate the eggs. One of them trembles. Soft peeping can be heard. Finally, a pink, wet chick pushes out from one end of the egg. The newborn female hawk is breathing hard. Little by little, she uncurls from the eggshell. The clumsy chick can't stand on her own yet, but she is here. She is hatched.

Hawks mate and nest each spring. Most hawks build nests high in tall trees. This gives them a good view of prey below. Rough-legged hawks and some red-tailed hawks build nests on cliffsides. A hawk's nest is big and messy-looking. It's made from sticks and sometimes bones. It has an inner cup lined with soft grass, feathers, and the fur of prey animals. Female hawks lay one to five eggs inside the inner cup.

In many species, hawks breed with the same partner each year. Pairs may use the same nest for many years. When they're not raising chicks, most hawks live alone. Some stay close to their nest tree all year.

3 WEEKS

- Grows quickly
- Weight: 1.6 pounds (714 g)
- Real feathers are growing in

5 WEEKS

- Weight: 2.2 pounds (980 g)
- Parents now drop food in nest
- Practices flapping

CLOSE-UP

Long tails

The long tail feathers of a true hawk act like a boat rudder. They help the hawk change direction quickly in midair. Since true hawks often live in forests, they must zigzag around tree trunks and other obstacles when they fly low to the ground.

FEATURED FAMILY

First Meal

The mother hawk brings a squirming gopher snake. She uses her beak to rip off small chunks of the animal. She passes the chunks from her beak to her chick's beak. The newborn chick is dry and fluffy now, but she is still uncoordinated. Her head bobbles. She drops her food, and her mother gives it to her again. Sometimes, the chick tumbles over while reaching for food. For the next six weeks, the hawk parents will bring meals for her and her siblings. As they grow, they will eat mice, squirrels, snakes, and other small creatures.

Hawks are apex predators. Adults are fierce hunters at the top of their food webs. In many places, hawks have a bad reputation. People don't like them because they can attack chickens and small pets. However, just like all creatures, hawks must find food. They must also protect their chicks from becoming another animal's meal. Eagles and other large raptors eat hawk eggs and chicks.

NORTH AMERICAN HAWKS

ACCIPITERS

These birds have short, wide, rounded wings and a long tail. They frequently flap their wings.

Examples:

Northern Goshawk
- Gray with pale underbody
- Found in forests

Cooper's Hawk
- Rust-colored underbody
- Found in forests, suburbs, and cities

Sharp-Shinned Hawk
- Rust-colored underbody
- Found in forests

BUTEOS

These birds have broad wings and a broad, rounded tail. They soar in wide circles high in the air.

Examples:

Red-Tailed Hawk
- Dark back and reddish-brown tail
- Found in open country

Rough-Legged Hawk
- Dark belly and white tail
- Found in cold climates

Swainson's Hawk
- Dark breast and light feathers
- Found in grasslands

CLOSE-UP

Color morphs

Different colors within a species are called **morphs**. The red-tailed hawk has three morphs. The light morph has a pale belly. The dark morph has a dark brown belly. The intermediate morph has a reddish-brown chest and a dark brown belly.

CHAPTER TWO

EARLY ADVENTURES

Young hawks stay in the nest for a month or two. As they grow and get stronger, they develop the skills they will need to fly. First, they crawl out of the nest onto a branch. They practice flapping their wings. In time, they flap hard enough to lift up into the air. They land right back on the same branch.

Next, they begin to take short hopping flights from branch to branch. Within a couple of weeks, they are ready to **fledge**, or leave the nest. The fledglings aren't ready to take care of themselves, though. For several more weeks, their parents continue to catch food for them.

6 WEEKS

- Takes first wobbly flight
- Leaves the nest
- Parents still provide food

9 WEEKS

- Catches own food
- Flying skills rapidly improve

CLOSE-UP

Feathered legs

Some hawks living in cold climates have feathers all the way down to their toes. The feathers provide insulation and help keep the birds warm.

FEATURED FAMILY

Look Who's Jump-Flapping!

After five weeks, the chick has lost her soft fuzzy coat. She's now covered in feathers. They're dark on her back and pale on her belly. Her head is white. She doesn't have the reddish-brown tail feathers of an adult red-tailed hawk. They won't grow in for another year. The chick can't fly yet, but she's learning the basic skills. She perches on the edge of the nest high in the saguaro. She flaps her wings. She's learning to balance, and her chest muscles are growing stronger. She even jumps in the air while flapping. Her body's getting ready to fly.

Young hawks face dangers while they learn to fly and hunt. Great horned owls are one threat. They regularly eat hawk chicks. Animals that hunt in trees or on the ground can also catch chicks. Weasels and wolves have been known to prey on young hawks.

Even after they can fly on their own, young hawks stay close to the nest. Being near their parents offers some protection. The mother and

3 MONTHS

- Improves hunting skills without help of parents
- Has reached full size
- Females are 25 percent larger than males
- Makes a short, shrill cry
- Parents chase off great horned owls and other predators

father also encourage their offspring to develop hunting skills. Gradually, the adults give less and less help with hunting. The young hawks begin to improve on their own. By the time they're several months old, most hawks are catching all their own food. Many will live in their parents' territory for up to a year.

CLOSE-UP

Bird bones

All birds have bones that contain lots of little air-filled holes. The bones are lightweight, making it easier for birds to take flight.

FEATURED FAMILY

Give It a Try

The young hawk is 44 days old. She clutches the sticks at the edge of the nest. The hot desert wind ruffles her feathers. She spreads her wings wide and flaps once or twice. Then she hops to a new spot. She flaps her wings again, gazing at the desert all around her. After a few more false starts, she spreads her wings again. This time, she pushes off, flapping hard. She sails through the blue sky. She flaps and flaps and finally...flies!

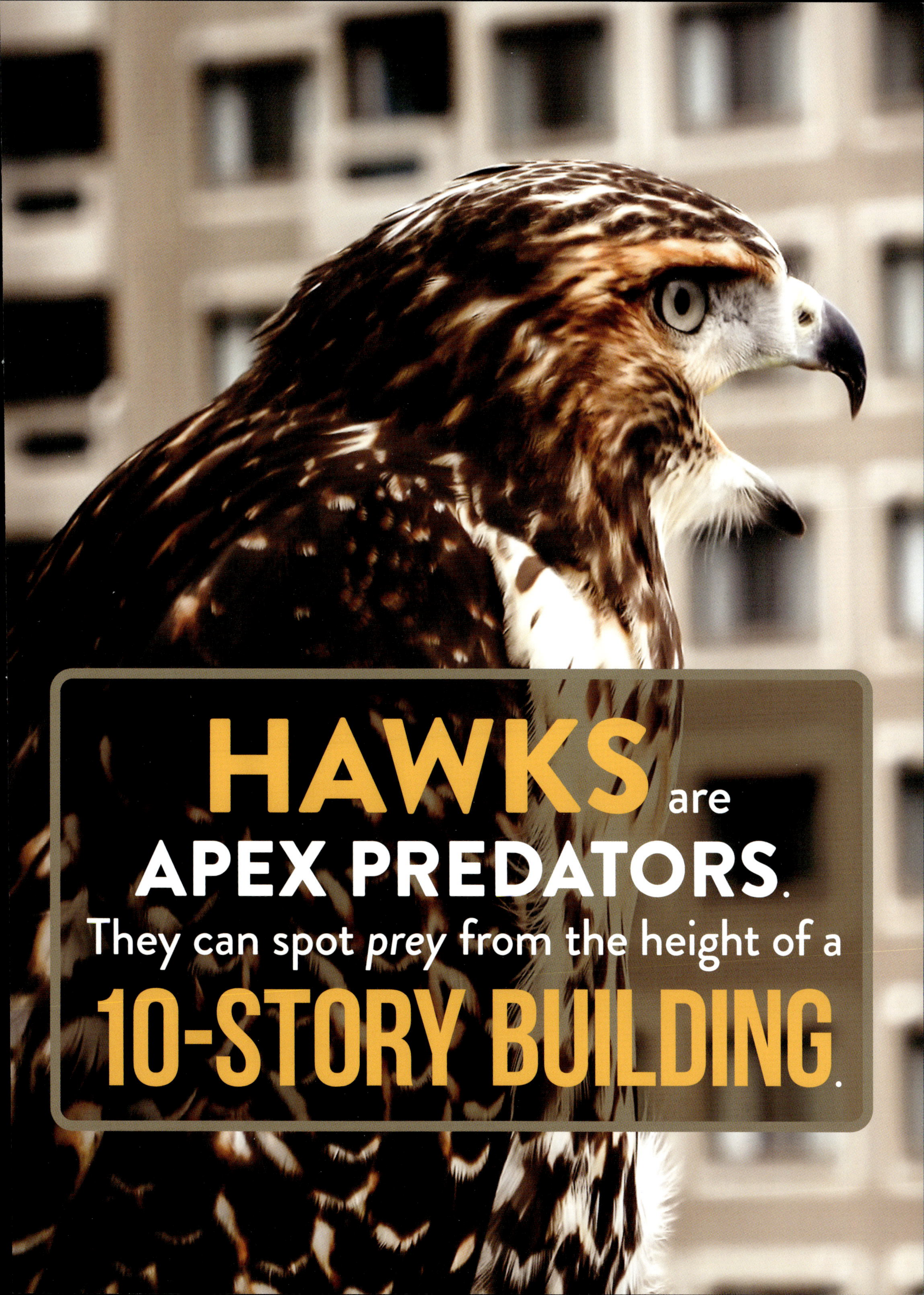

HAWKS are **APEX PREDATORS.** They can spot *prey* from the height of a **10-STORY BUILDING.**

CLOSE-UP

Kiting

Some hawks can kite. In this maneuver, a hawk holds still while flying as the wind blows around it. The hawk doesn't even have to flap its wings to keep itself aloft. It looks like a kite at the end of a string.

CHAPTER THREE

LIFE LESSONS

Hawks live all over the world. Different species live in many different habitats. Hawks live in forests and deserts. They live on the tundra and in mountain ranges. They live on islands and even in cities. They are most common in **tropical** areas of the planet.

A hawk's habitat affects its hunting style. The northern goshawk, a true hawk, often hunts at the edge of a forest. It flies low along the ground. This startles prey animals and gets them moving. Once the prey is on the run, the goshawk chases it down. It will even crash into thick shrubs to pursue prey. A long tail helps the goshawk change directions easily when flying through the forest and seeking out small, quick animals such as squirrels, rabbits, foxes, and mice.

Red-tailed hawks, which are buteos, have a different hunting style. These large birds have a wingspan of almost four feet (1 m), and they

1 YEAR

- Has become an expert, powerful flier
- Leaves parents to find own territory

2 YEARS

- Adapts hunting techniques for prey in different locations and seasons

can weigh four pounds (2 kg). Broad wings allow them to soar high in the air for long periods of time. They often circle over open fields. From this vantage point, they can see everything. Prey cannot hide. When a red-tailed hawk spots a tasty animal, it dives. As it descends, it extends its talons to grab the prey. The hawk usually carries the animal back to a perch to kill and eat it. In this way, red-tailed hawks hunt for voles, mice, rats, squirrels, rabbits, and other small mammals.

True hawks live for about 15 years in the wild. Buteos can live to be around 20 years old. Hawks grow quickly. They are fully grown when they are one to three months old. That doesn't mean their hunting skills are fully developed, though. That takes months and years of practice.

CLOSE-UP

Hunting injuries

True hawks often fly through trees and shrubs while hunting. This results in some battle scars. In one study, researchers found that one out of four skeletons of Cooper's hawks showed signs of previously broken chest bones.

FEATURED FAMILY

This Is How It's Done

The young hawk is able to leave the nest, but her parents still feed her. When they drop a dead mouse on the ground, she dives down to take it. They begin dropping food farther away. To get it, the fledgling must fly farther. Eventually, the parents drop a rabbit that is injured, but still alive. For the first time, the young hawk kills her food. For the next few weeks, with her parents' help, she practices eating in this way. Her hunting skills develop naturally. Soon, she is a predator. By August, she's catching all her own food.

3 YEARS

- Sexually mature
- Performs aerial displays, including soaring and diving, to attract a mate
- Male and female pair work together for about a week to build a nest
- Female keeps the eggs warm while the male hunts

Hawks reach sexual maturity at one to two years old. They usually have just one mate, and they breed once each year. When they're not raising young, most hawks live alone. However, mated pairs may stay in the same territory year-round.

CLOSE-UP

Back stacking

Usually, hawks perch on trees alone. But Harris's hawks are "back stackers." When there aren't enough good perches for everyone, they will stand on each other's backs.

Hawks are known for being fierce killers. These impressive birds are dangerous for small animals, but they are important for humans. They help farmers by hunting rodents. Mice and other pests damage farms and orchards by tearing open bags of grain and stripping the bark off fruit trees. Hawks and other raptors help keep rodent populations under control. Hawks are also magnificent animals to watch and learn about. Crowds gather to view hawks migrating in the spring and fall. People who practice the ancient sport of falconry train hawks to hunt on command.

FEATURED FAMILY

Practice Makes Perfect

The red-tailed hawk is now a skilled hunter. Sometimes, she glides high above the desert as she uses her keen eyesight to search for prey. Today, though, she simply perches at the top of a tall cactus and searches the landscape. When she spots a ground squirrel, she swoops down and grabs it with her talons. She has no teeth, so she cannot chew. Instead, she uses her sharp beak to rip the squirrel into pieces. She swallows each piece whole.

BEFORE THEY MATE, RED-TAILED HAWKS lock TALONS and SPIRAL DOWN TOGETHER through the AIR.

(20) YEARS

- Has outlived two mates
- Nests each year from March to May
- Has raised 17 broods of chicks
- End of life

CHAPTER FOUR

HELPING HAWKS SURVIVE

Hawks, like all modern birds, have dinosaur ancestors. Hawks have been on Earth for many millions of years.

Most hawks are surviving well, but some species are in trouble. The Gundlach's hawk of Cuba is endangered. The New Britain sparrowhawk of Papua New Guinea is vulnerable. Both species are true hawks that live in wooded areas. Cutting down forests is a big threat to these hawks. Widespread logging and mining may leave them with nowhere to nest.

Buteos don't depend quite as much on forests, so most buteo populations are stable or even increasing. However, one critically endangered species is the Ridgway's hawk. It lives in the Dominican Republic and Haiti. Loss of its habitat is a major problem. People clear land in order to farm. They see the hawk as a threat to chickens on their farms. They take down the hawk's nests and destroy its eggs.

Other threats to this bird include long periods of rain that collapse nests. The larvae of some flies live off the blood of Ridgway's hawk chicks. They kill many nestlings.

Conservation groups around the world work to support hawks and other raptors. HawkWatch International studies raptors, manages migration sites, and educates students. The Hawk Conservancy Trust also works to save raptors. Its members research buteos in Great Britain. They study the effects of **pesticides** on these birds. They track buteos to see what factors affect their populations. They run the National Bird of Prey Hospital, which treats more than 200 raptors each year for injuries and illnesses. The hospital also houses raptor breeding programs.

Hawk Mountain Sanctuary was the world's first raptor refuge. It opened in Pennsylvania in 1935 on land where raptors were frequently shot. This large refuge allows visitors to learn about raptors in their natural habitat. The sanctuary trains scientists and teachers. People who studied there have gone on to open similar raptor refuges in the Arctic, Central America, and South America.

Hawks are vital to the health of our wilderness areas, and sometimes they are citizens of our cities. These strong, fierce creatures deserve their place in our world.

FAMILY ALBUM

SNAPSHOTS

Red-tailed hawks have a distinctive cry. When a movie features a bald eagle, the director almost always chooses to include a recording of a red-tailed hawk's cry instead of the eagle's cry.

Ferruginous hawks live in the wide-open spaces of the western United States. They thrive in grasslands and on prairies.

Tiny **sharp-shinned hawks** weigh only three to seven ounces (85 to 198 g). They zip through dense forests to prey on songbirds.

Rough-legged hawks in the Arctic sometimes use caribou bones to build their nests.

The **northern goshawk** strikes its prey at a speed of about 50 miles (80 kilometers) per hour.

Swainson's hawks often migrate in huge groups with other bird species. Groups may include tens of thousands of birds.

The **Cooper's hawk** often visits bird feeders to snack on songbirds.

The chicks of **red-shouldered hawks** shoot their poop over the edge of the nest. This helps keep the nest clean and tidy.

The **Hawaiian hawk** was considered a symbol of royalty in traditional Hawaiian culture.

Gray hawks live mostly in Mexico and Central America. Their diet is mostly lizards.

A group of **Harris's hawks** will surround prey and take turns chasing it.

WORDS to Know

accipiters hawks with short, broad wings; accipiters live mainly in forests and are also known as true hawks

apex predators animals at the top of the food chain; animals that hunt and that are not usually hunted by other animals

buteos large birds of prey that are called hawks in the United States but that are known as buzzards in other parts of the world; buteos live mainly in open areas

fledge to leave the nest after growing flight feathers

morphs members of a species who share a particular coloring or pattern

pesticides chemicals used on farms and in other areas to kill pests, especially insects

raptors birds of prey with sharp beaks and talons; eagles, owls, and hawks are all raptors

species a group of living things that have shared characteristics and that are able to reproduce with one another

talons the claws of a bird of prey

tropical having warm weather all year

LEARN MORE

Books

Brereton, Catherine. *Nature Guide: Birds*. London: Bloomsbury Children's Books, 2020.

Hoena, Blake. *Everything Birds of Prey*. Washington, D.C.: National Geographic, 2015.

Riggs, Kate. *Hawks*. Mankato, Minn.: Creative Education/Creative Paperbacks, 2022.

Websites

"Cornell Lab of Ornithology." Cornell University.
https://www.birds.cornell.edu/home

"Red-Tailed Hawks Live Cam." Cornell Lab of Ornithology.
https://www.allaboutbirds.org/cams/red-tailed-hawks/

"The Raptor Center." University of Minnesota.
https://raptor.umn.edu/about-raptors/learn-about-raptors

Documentaries

Attenborough, David. *Redtail: The Story of a Hawk*. BBC Wildlife on One, 1985.

Hess, Janet, and Frederic Lilien. *Pale Male*. PBS Nature, 2004.

Lange, Willem. *Journey of the Broad-Winged Hawk*. New Hampshire PBS, 2010.

Visit

CALIFORNIA RAPTOR CENTER

Tour the center and meet many injured and healing raptors.

University of California, Davis
1340 Equine Lane
Davis, CA 95616

HAWK MOUNTAIN SANCTUARY

Join a program or take a hike at this wildlife sanctuary.

1700 Hawk Mountain Rd.
Kempton, PA 19529

THE CENTER FOR BIRDS OF PREY

Tour the center and see raptors fly.

Avian Conservation Center
4719 Highway 17N
Awendaw, SC 29429

THE RAPTOR CENTER

Take a tour to see and learn about raptors.

University of Minnesota
1920 Fitch Avenue
St. Paul, MN 55108

INDEX

beaks, 6–8, 10, 22
buteos, 7, 8, 11, 19, 20, 25, 26
chicks, 8–11, 14, 15, 23, 26, 29
colors, 7, 11, 12
conservation, 26
eggs, 4, 7–9, 11, 21, 25
feathers, 7, 9–11, 14, 16
flights, 13, 16
forests, 8, 10, 11, 19, 25, 28
habitats, 19, 25, 26
hunting, 15, 16, 19–22
nests, 4, 8, 9, 13–16, 20, 21, 23, 25, 26, 28, 29
pairs, 4, 9, 21, 22
predators, 11, 15, 17, 20
prey, 7–9, 15, 17, 19, 20, 22, 28, 29
raptors, 7, 11, 22, 26
rodents, 22
tails, 7, 10, 11, 14, 19
talons, 4, 7, 8, 20, 22, 23
wings, 7, 11, 13, 14, 16, 18, 20